THE OFFICIAL SASSAFRAS SCIDAT LOGBOOK

BOTANY EDITION

PREPARED BY: _____

The Official Sassafras SCIDAT Logbook: Botany Edition

Third Edition, First Printing 2022
Copyright @ Elemental Science, Inc.
Email: support@elementalscience.com

ISBN # 978-1-935614-31-9

Printed In USA For World Wide Distribution

For more copies write to :
Elemental Science
PO Box 79
Niceville, FL 32588
support@elementalscience.com

Copyright Policy

The Official Sassafras SCIDAT Logbook:
Botany Edition

Table of Contents

SCIDAT* LOGBOOK AT-A-GLANCE

Document your adventure with the Sassafras Twins!

1. INFORMATION SHEET

The purpose of this sheet is for you to record what you have learned about the various divisions of the plant kingdom studied in *The Sassafras Science Adventures Volume 3: Botany.*

2. BIOME SHEET

The purpose of this sheet is to give you an opportunity to work on your mapping skills as you study the different biomes around the world.

3. BOTANY RECORD SHEET

The purpose of this sheet is for you to record what you have learned about the plants that are introduced in *The Sassafras Science Adventures Volume 3: Botany.*

4. PROJECT RECORD SHEET

The purpose of this sheet is for you to record the projects you have completed throughout the course of your study of botany.

5. BOTANY NOTES SHEET

The purpose of this sheet is for you to record any additional information you have learned during your study of botany.

6. BOTANY GLOSSARY

The purpose of the glossary is for you to create a glossary of vocabulary terms you have encountered throughout your reading of *The Sassafras Science Adventures Volume 3: Botany.*

*SCIDAT stands for "**sci**entific **dat**a" and it comes from the app that the Sassafras Twins use on their journey.

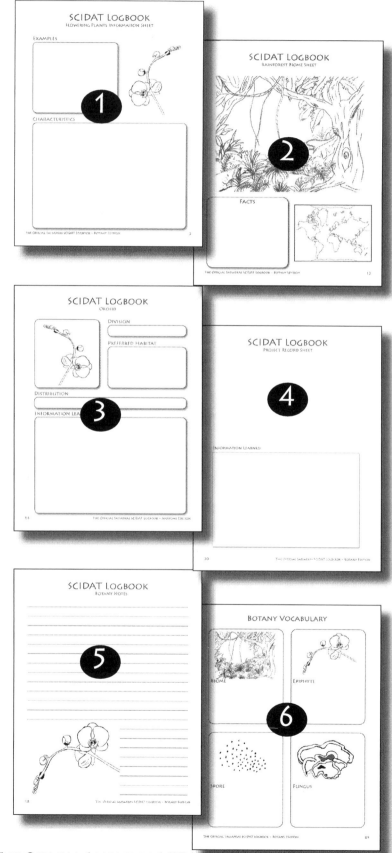

SCIDAT Logbook
Flowering Plants Information Sheet

Examples

Characteristics

SCIDAT Logbook
Herbs and Grasses Information Sheet

Examples

Characteristics

SCIDAT Logbook

Carnivorous and Parasitic Plants Information Sheet

Examples

Characteristics

SCIDAT Logbook
Non-flowering Plants Information Sheet

Examples

Characteristics

SCIDAT Logbook
Fungi and Molds Information Sheet

Examples

Characteristics

SCIDAT Logbook
Trees Information Sheet

Flowering Examples

Non-flowering Examples

Characteristics

SCIDAT Logbook
Find Your Habitat

WHAT KINDS OF PLANTS DO YOU SEE?

_____

_____

_____

_____

WHAT KINDS OF ANIMALS DO YOU SEE?

_____

_____

_____

_____

WHAT IS THE WEATHER LIKE TODAY?

_____

_____

_____

_____

WHAT IS THE WEATHER USUALLY LIKE IN THE DIFFERENT SEASONS?

_____

_____

_____

_____

SCIDAT Logbook

_____
_____
_____
_____
_____
_____
_____
_____
_____
_____
_____
_____
_____
_____
_____
_____
_____
_____

SCIDAT Logbook
Rainforest Biome Sheet

Facts

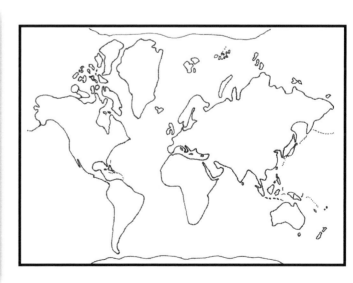

SCIDAT Logbook
Orchid

Division

Preferred Habitat

Distribution

Information Learned

SCIDAT Logbook
Fern

Division

Preferred Habitat

Distribution

Information Learned

SCIDAT Logbook
Peace Lily

Division

Preferred Habitat

Distribution

Information Learned

SCIDAT Logbook
Shelf Fungus

Division

Preferred Habitat

Distribution

Information Learned

SCIDAT Logbook

SCIDAT Logbook
Botany Notes: Amazon Rainforest

_____

_____

_____

_____

_____

_____

_____

_____

_____

_____

_____

_____

_____

_____

_____

_____

_____

_____

SCIDAT Logbook
Project Record Sheet

Glue Picture of
Project Here

Information Learned

SCIDAT Logbook
Project Record Sheet

Glue Picture of
Project Here

Information Learned

SCIDAT Logbook
Roses

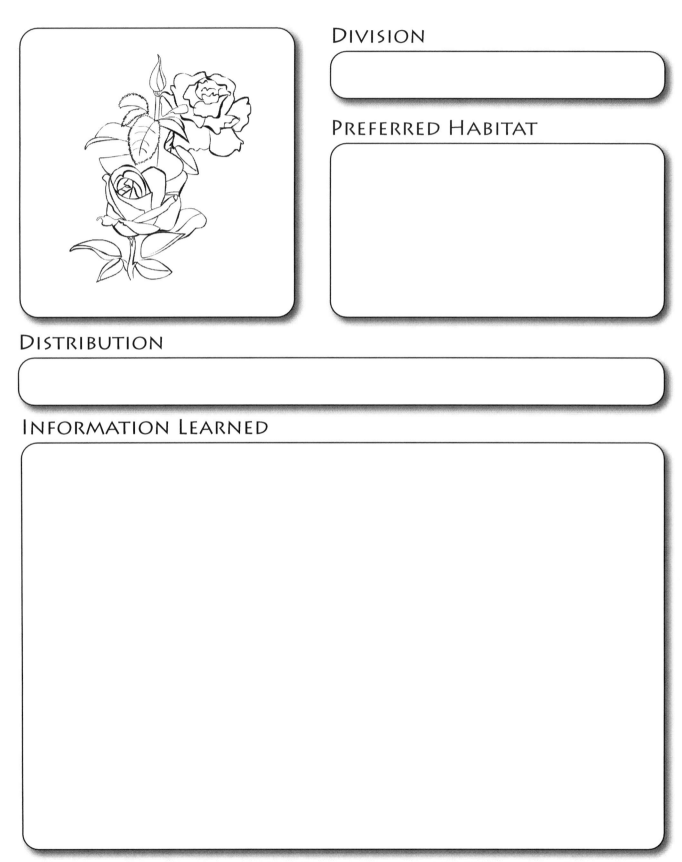

Division

Preferred Habitat

Distribution

Information Learned

SCIDAT Logbook
Hedges

Division

Preferred Habitat

Distribution

Information Learned

SCIDAT Logbook
POLLINATION

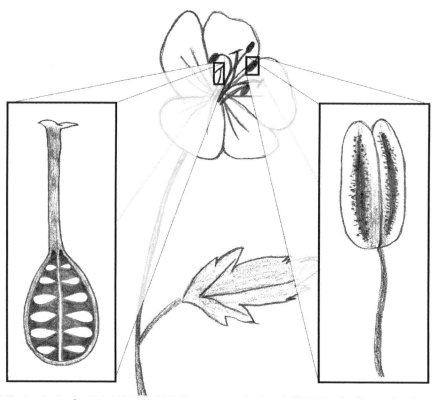

INFORMATION LEARNED

SCIDAT Logbook
Peat Bog Biome Sheet

Facts

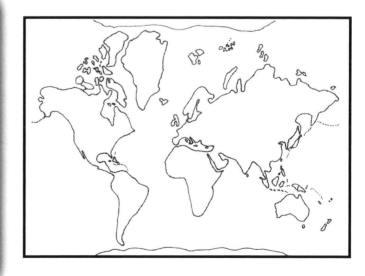

SCIDAT Logbook
Moss

Division

Preferred Habitat

Distribution

Information Learned

SCIDAT Logbook
Peat Bog

Typical Plants

Distribution

Information Learned

SCIDAT Logbook
BOTANY NOTES: SCOTTISH GARDEN AND PEAT BOG

SCIDAT Logbook

Botany Notes: Scottish Garden and Peat Bog

SCIDAT Logbook
Project Record Sheet

Glue Picture of
Project Here

Information Learned

SCIDAT Logbook
Project Record Sheet

Glue Picture of
Project Here

Information Learned

SCIDAT Logbook
Grasslands Biome Sheet

Facts

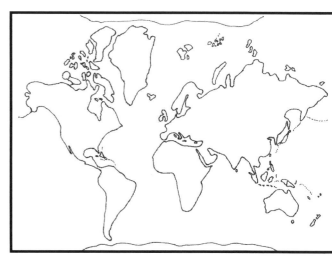

SCIDAT Logbook
The Pampas

Typical Plants

Distribution

Information Learned

SCIDAT Logbook
Pampas Grass

Division

Preferred Habitat

Distribution

Information Learned

SCIDAT Logbook
Verbena

Division

Preferred Habitat

Distribution

Information Learned

SCIDAT Logbook

Ombú

Division

Preferred Habitat

Distribution

Information Learned

SCIDAT Logbook
Botany Notes: Argentinian Pampas

_____

_____

_____

_____

_____

_____

_____

_____

_____

_____

                             _____

                             _____

                             _____

                             _____

                             _____

                             _____

_____

SCIDAT Logbook

SCIDAT Logbook
Project Record Sheet

GLUE PICTURE OF
PROJECT HERE

INFORMATION LEARNED

SCIDAT Logbook
Project Record Sheet

Glue Picture of
Project Here

Information Learned

SCIDAT Logbook
Palm

Division

Preferred Habitat

Distribution

Information Learned

SCIDAT Logbook
Pitcher Plant

Division

Preferred Habitat

Distribution

Information Learned

SCIDAT Logbook
Mold

Division

Preferred Habitat

Distribution

Information Learned

SCIDAT Logbook
GIANT Rafflesia

Division

Preferred Habitat

Distribution

Information Learned

SCIDAT Logbook

Botany Notes: Bornean Beachside and Rainforest

SCIDAT Logbook

Botany Notes: Bornean Beachside and Rainforest

_____

_____

_____

_____

_____

_____

_____

_____

_____

_____

_____

_____

_____

_____

_____

_____

_____

_____

_____

_____

_____

SCIDAT LOGBOOK
PROJECT RECORD SHEET

GLUE PICTURE OF
PROJECT HERE

INFORMATION LEARNED

SCIDAT Logbook
Project Record Sheet

GLUE PICTURE OF
PROJECT HERE

INFORMATION LEARNED

SCIDAT Logbook
Taiga Biome Sheet

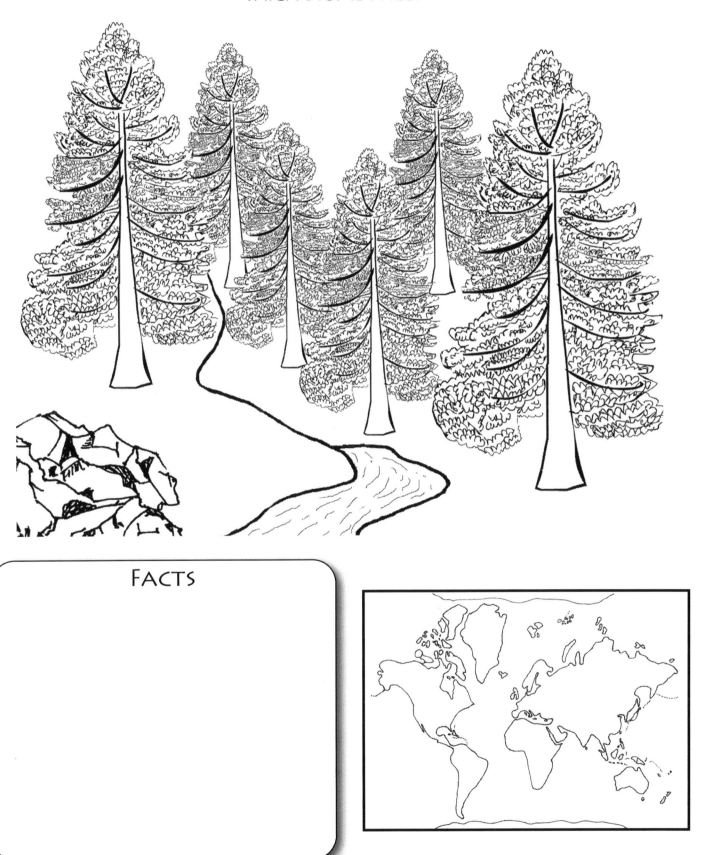

Facts

SCIDAT Logbook
Dwarf Birch Shrub

Division

Preferred Habitat

Distribution

Information Learned

SCIDAT Logbook
Crocus

Division

Preferred Habitat

Distribution

Information Learned

SCIDAT Logbook
Tundra Biome Sheet

Facts

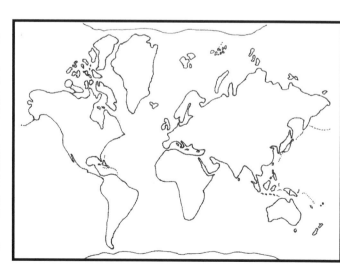

SCIDAT Logbook
Lichen

Division

Preferred Habitat

Distribution

Information Learned

SCIDAT Logbook
Algae

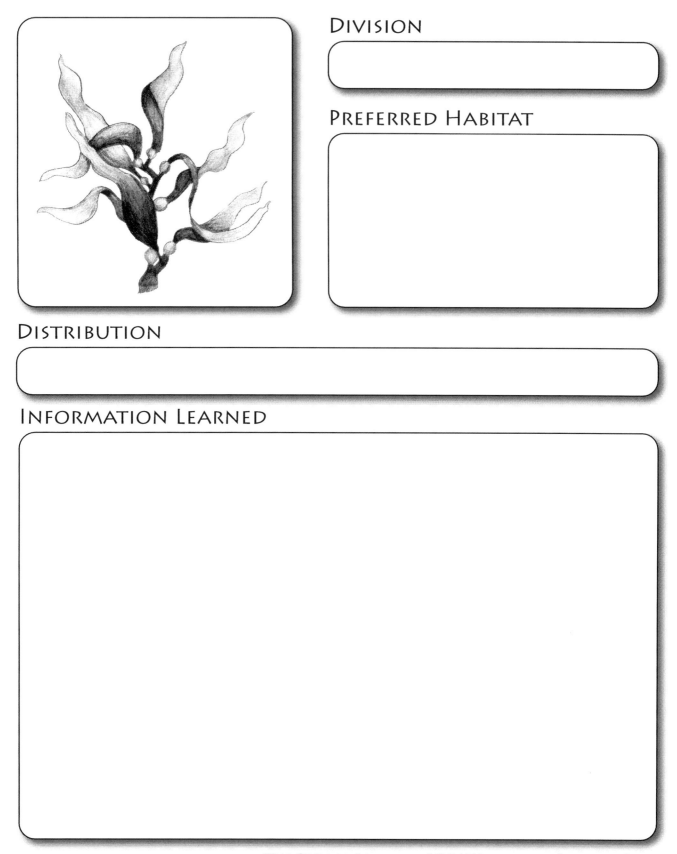

Division

Preferred Habitat

Distribution

Information Learned

SCIDAT Logbook
Botany Notes: Siberian Taiga and Tundra

_____

_____

_____

_____

_____

_____

_____

_____

_____

_____

_____

_____

                        _____

                        _____

                        _____

                        _____

                        _____

                        _____

                        _____

SCIDAT Logbook

Botany Notes: Siberian Taiga and Tundra

_____

_____

_____

_____

_____

_____

_____

_____

_____

_____

_____

_____

_____

_____

_____

_____

_____

_____

_____

_____

SCIDAT Logbook
Project Record Sheet

GLUE PICTURE OF
PROJECT HERE

INFORMATION LEARNED

SCIDAT Logbook
Project Record Sheet

Glue Picture of
Project Here

Information Learned

SCIDAT Logbook
Plant Cells

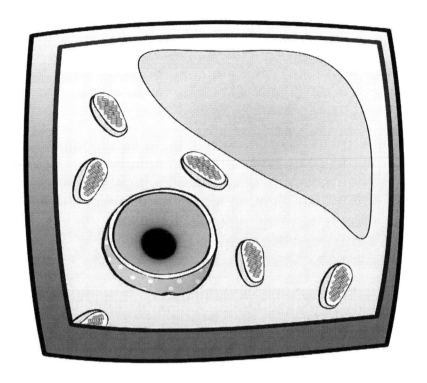

Information Learned

SCIDAT Logbook
Photosynthesis

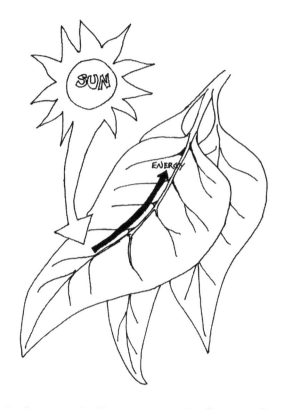

INFORMATION LEARNED

SCIDAT Logbook
Deciduous Forest Biome Sheet

Facts

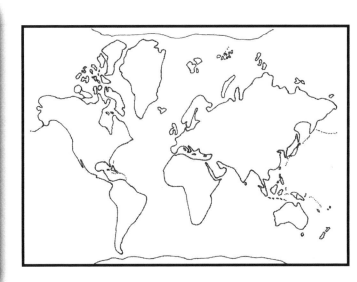

SCIDAT Logbook
Chestnut Tree

Division

Preferred Habitat

Distribution

Information Learned

SCIDAT Logbook
Apple Tree

Division

Preferred Habitat

Distribution

Information Learned

SCIDAT Logbook
Botany Notes: French Deciduous Forest

SCIDAT Logbook
Botany Notes: French Deciduous Forest

SCIDAT Logbook
Project Record Sheet

Glue Picture of
Project Here

Information Learned

SCIDAT Logbook
Project Record Sheet

Glue Picture of
Project Here

Information Learned

SCIDAT Logbook
Sitka Spruce

Division

Preferred Habitat

Distribution

Information Learned

SCIDAT Logbook
Cone

Description

Information Learned

SCIDAT Logbook
Redwood

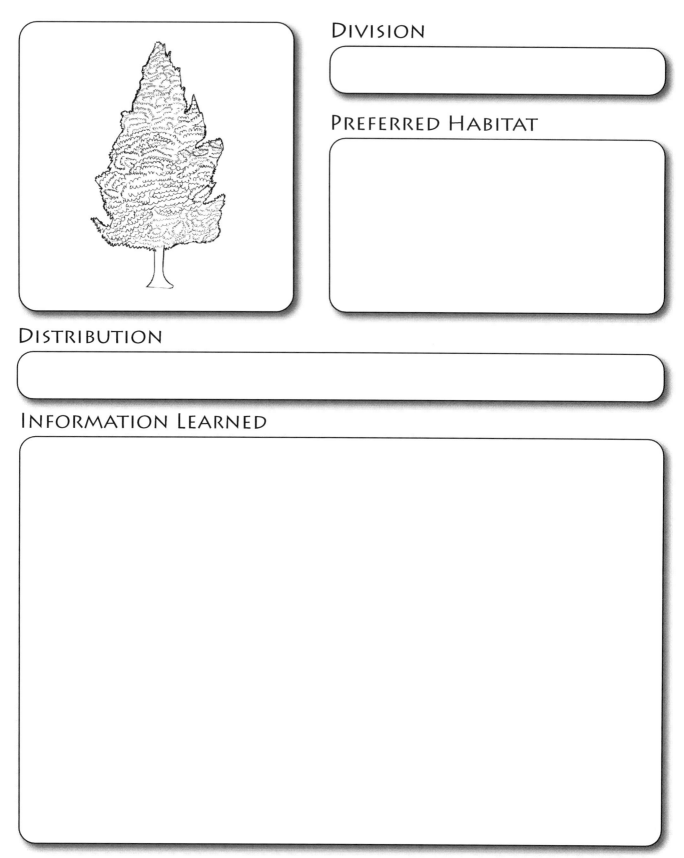

Division

Preferred Habitat

Distribution

Information Learned

SCIDAT Logbook

Mushroom

Division

Preferred Habitat

Distribution

Information Learned

SCIDAT Logbook

Botany Notes: Pacific Coastal Forest

SCIDAT Logbook
Botany Notes: Pacific Coastal Forest

SCIDAT Logbook
Project Record Sheet

Glue Picture of
Project Here

Information Learned

SCIDAT Logbook
Project Record Sheet

Glue Picture of
Project Here

Information Learned

SCIDAT Logbook
Desert Biome Sheet

Facts

SCIDAT Logbook
Joshua tree

Division

Preferred Habitat

Distribution

Information Learned

SCIDAT LOGBOOK
BARREL CACTUS

DIVISION

PREFERRED HABITAT

DISTRIBUTION

INFORMATION LEARNED

SCIDAT Logbook
Creosote Bush

Division

Preferred Habitat

Distribution

Information Learned

SCIDAT Logbook
Paddle Cactus

Division

Preferred Habitat

Distribution

Information Learned

SCIDAT Logbook
Botany Notes: Mojave Desert

_____

_____

_____

_____

_____

_____

_____

_____

_____

_____

_____

_____

_____

_____

_____

_____

SCIDAT LOGBOOK
BOTANY NOTES: MOJAVE DESERT

_____

_____

_____

_____

_____

_____

_____

_____

_____

_____

_____

_____

_____

_____

_____

_____

_____

_____

_____

_____

_____

SCIDAT Logbook
Project Record Sheet

Glue Picture of
Project Here

Information Learned

SCIDAT Logbook
Project Record Sheet

Glue Picture of
Project Here

Information Learned

SCIDAT Logbook
Botany Notes: Bonus Data

BONUS DATA

WETLANDS

Wetlands are saturated areas of land also known as marshes or swamps. There are salt and freshwater marshes, which are full of plants and animals that have adapted to an environment with an abundance of water. Typically, wetlands are found in coastal flats alongside a body of water or in inland depressions. This is because in these areas the water cannot drain away quickly. However, the water does move slowly through these areas, so it is possible for some parts of the swamp to dry out completely.

SCIDAT Logbook
Botany Notes: Bonus Data

_____

_____

_____

_____

_____

_____

_____

_____

_____

_____

_____

_____

_____

_____

_____

BONUS DATA

SWAMP PLANTS

Plants in a swamp vary depending upon location, but generally, swamps can be divided into two groups based on the plants the swamps contain.

Shrub swamps are dominated by shrubs, such as the dogwood and mangrove. Forested swamps are dominated by trees, such as cypress, willow, and red maple. These trees have developed specialized underwater roots that can trap sediment, which allows the tree to build its own stable base in a watery environment.

BOTANY GLOSSARY

Botany Vocabulary

Biome

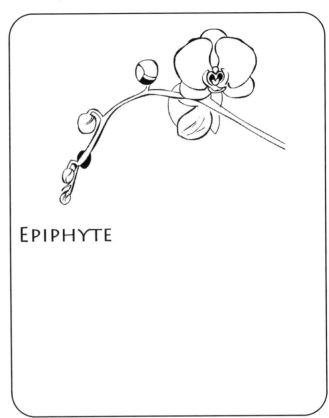

Epiphyte

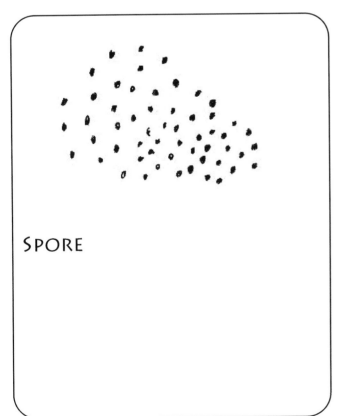

Spore

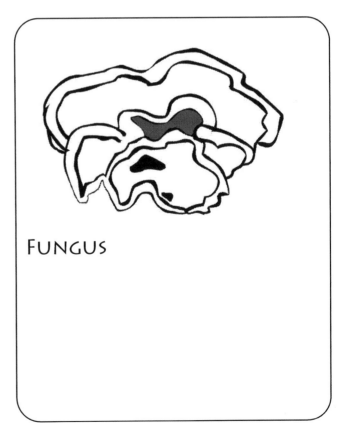

Fungus

Botany Vocabulary

ANGIOSPERM

FLOWER

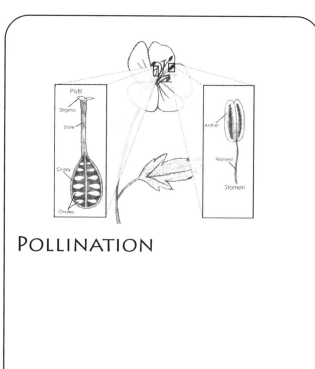

POLLINATION

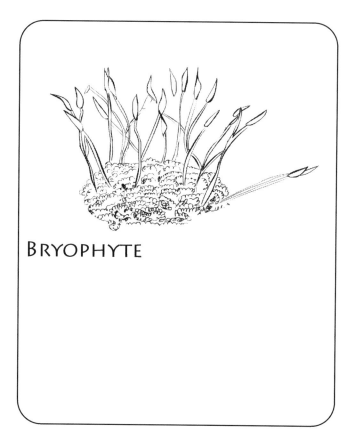

BRYOPHYTE

Botany Vocabulary

GRASSES

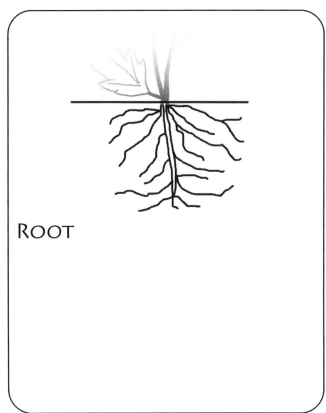

ROOT

WILDFLOWER

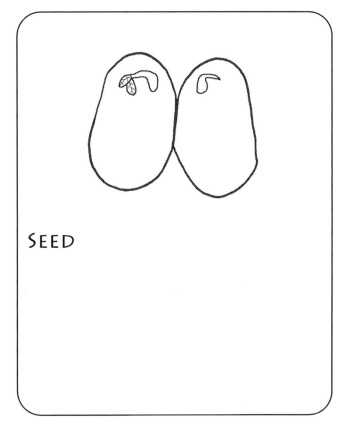

SEED

Botany Vocabulary

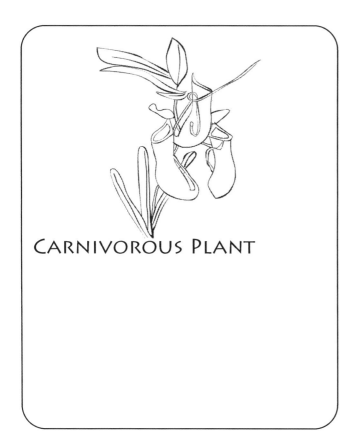

Carnivorous Plant

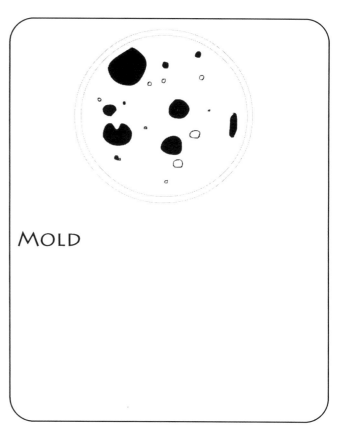

Mold

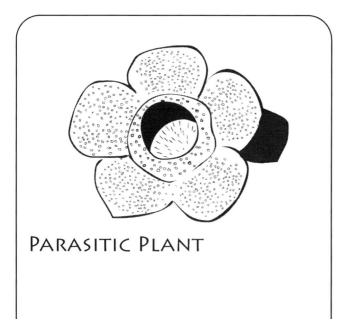

Parasitic Plant

Taiga

Botany Vocabulary

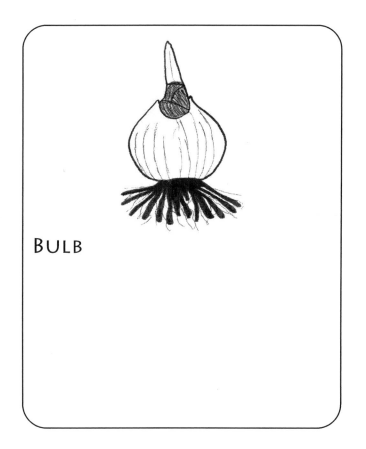

BULB

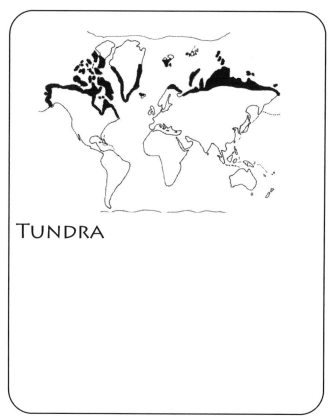

TUNDRA

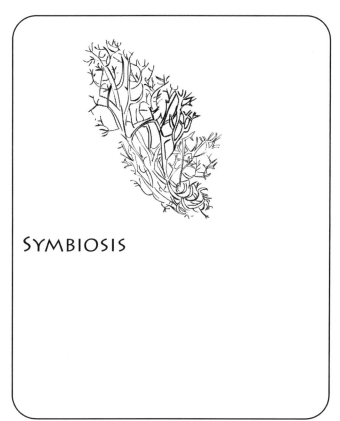

SYMBIOSIS

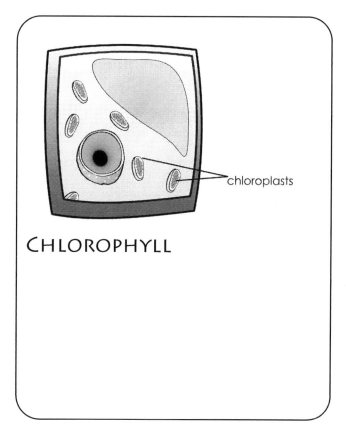

chloroplasts

CHLOROPHYLL

Botany Vocabulary

Leaf

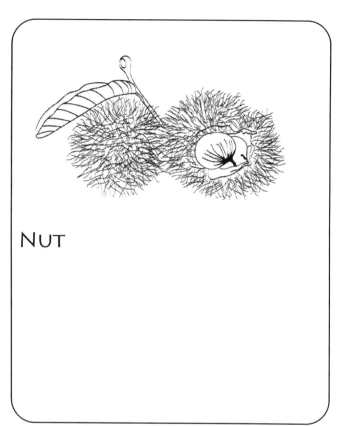

Nut

Deciduous

Fruit

Botany Vocabulary

CONIFER

CONE

GYMNOSPERM

MUSHROOM

BOTANY VOCABULARY

SUCCULENT

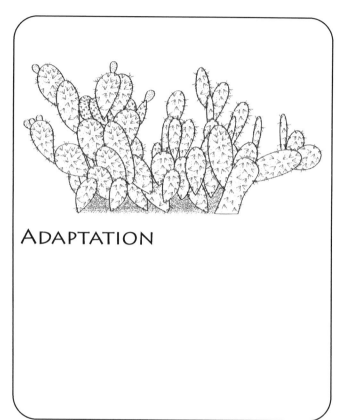

ADAPTATION